MRSA and Staphylococcal Infections

by

Hernan R. Chang, M.D.

Also by Hernan R. Chang

Elysium: A Collection of Haiku and Senryu

MRSA- Spider Bites: The Flesh-Eating Bacterial Epidemic that Threatens America

Copyright © 2006 by Hernan R. Chang
All Rights Reserved

No part of this book may be reproduced or transmitted in any form or by any means –graphic, electronic, or mechanical– including photocopying, recording, taping or by any information storage or retrieval system, without permission in writing from the author. It is sold with the understanding that the author is not engaged in rendering medical, health or any other kind of personal professional services in the book. Readers are advised to seek medical help when appropriate. It is the responsibility of a treating physician to determine medicine dosage and the best treatment for a given patient. The patient vignettes are the product of the author's imagination. Any resemblance to a real patient is purely coincidental. The naming of antibiotics, medicines, procedures and tests in this book does not constitute endorsement for their use. Photographs were taken only after full consent was obtained from the patients. The author disclaims all responsibility for any liability, loss, or risk, personal or otherwise, which is incurred as a consequence, directly or indirectly, of the use and application of any of the contents of this book.

ISBN-13: 978-1-84728-327-6

Printed in the United States of America

We always deserve what we tolerate,
but we never tolerate what we deserve.

Table of Contents

Foreword	iii
Chapter I	1
What are Staph and MRSA?	1
Chapter II	7
How do we get Staph and/or MRSA infections?	7
Chapter III	15
What diseases can Staph and MRSA produce?	15
Chapter IV	19
How do we diagnose a Staph or MRSA infection?	19
Chapter V	21
How do we treat Staph and MRSA infections?	21
Chapter VI	23
How do we prevent infections with Staph and MRSA?	23
Chapter VII	29
Patient Vignette 1	29
Patient Vignette 2	34
Chapter VIII	45
Conclusions	45
Photographs	47

Appendix A	57
Handwashing	57
Appendix B	59
Mupirocin for Nasal Decolonization	59
Appendix C	61
Chlorhexidine for Skin Decolonization	61
Bibliography	63
Glossary	73
Index	79

Foreword

We are witnessing a worldwide increase in infections with *Staphylococcus aureus*, commonly called *S. aureus*, methicillin-sensitive *S. aureus* (MSSA) or simply "Staph." Many of these infections are due to invasive strains of methicillin-resistant *S. aureus* (MRSA). These infections are responsible for causing significant morbidity and burden upon healthcare systems.

People not involved in a healthcare setting are often not aware of this silent epidemic. This lack of awareness includes the significance and danger of "community-acquired" MRSA infections and their potential threat to the healthcare system. The contents of this book are geared to the reader who does not have a medical or healthcare industry background. The information presented is intended to provide a general understanding about Staph infections. In order to curb the worldwide menace of Staph and MRSA infections, a significant behavioral change is needed. A well-informed public is essential for this behavioral change to occur. The opinions presented in this book are the author's and are not intended to replace the professional advice of a healthcare provider which would be given after a careful evaluation and clinical examination. This book is formatted in large print for the visually-impaired reader.

Chapter I
What are Staph and MRSA?

What is Staph?

Staph (also known as *Staphylococcus aureus* or *S. aureus*) is a bacterium that is commonly found in the environment and on the skins, noses, armpits, and groins of healthy people. This bacterium can cause illnesses ranging from minor skin infections (like pimples or boils) to life-threatening diseases such as Toxic Shock Syndrome. About one-third of the United States population carries Staph in their noses and may or may not have any symptoms or disease. These people are said to be "colonized" by Staph.

How common are Staph infections?

Staph is one of the most common bacteria producing skin infections in the United States and the world. These skin infections range from minor ailments, such as boils or other skin conditions, to severe cases such as pneumonia, bacteremia, toxic shock syndrome, and "flesh-eating" disease (i.e., necrotizing fasciitis).

Why do some people become "colonized" with Staph, but never develop symptoms or disease?

People who are colonized with Staph have the organism living on or in the body, but do not have any signs or symptoms of illness or infection. They also may be colonized in multiple sites of the body, but again with no signs of illness. Why is this? One reason may be due to an equilibrium that exists between the bacteria present in the nose or other body parts and the ability of the body to fight the bacteria. In addition, many other types of bacteria live on the skin along with Staph. These other bacteria take up space on the body and do not allow the Staph to completely take over. A strong immune system may also contribute to the prevention of Staph invasion to deeper areas of the body. *It is important to remember, however, that regardless of whether a patient is colonized or infected with Staph, the organism can still be transmitted to another person, primarily through skin-to-skin and hand contact.*

How do people who are colonized with Staph develop a Staph infection?

Anyone colonized with Staph can develop a Staph infection later on in her or his life. How? The bacteria are generally harmless unless they enter the body through a wound or other cut, and even then, they usually only cause minor skin problems in healthy people. The bacteria can become deadly, however, when they enter the bodies of people who are already ill or have weakened immune systems.

What is MRSA?

The common Staph is also called "methicillin-sensitive" *S. aureus* or MSSA. MRSA stands for "methicillin-resistant" *Staphylococcus aureus*, which is a strain of Staph that emerged in hospitals many decades ago. Typically, when a person gets a Staph infection, he/she is treated with antibiotics commonly used to treat skin infections. This new strain, however, was found to be resistant to an antibiotic called methicillin, which is why the new strain is called "methicillin-resistant" *Staphylococcus aureus* (MRSA). MRSA is also resistant to other antibiotics such as oxacillin, nafcillin, penicillin, ampicillin, amoxicillin and a group of antibiotics called cephalosporins. There are other antibiotics, however, that can be used to treat MRSA infections such as vancomycin, linezolid, daptomycin, tigecycline and quinupristin-dalfopristin. Although approximately one-third of the US population is colonized with Staph, only a minority of these are MRSA.

How common are MRSA infections?

Back in the 1960s, MRSA was found in hospital settings among elderly patients and patients residing in nursing homes. However, in the last few years, a new type of MRSA (having the Panton-Valentine toxin) has emerged affecting people of all ages, many of whom have never been in a hospital! This new type of MRSA, originally found among wrestlers, football players, inmates and military recruits, has now been found to be the most common cause of skin and soft-tissue infections among patients treated in emergency rooms across the United States, according to a recent study.

How frequent are colonizations with Staph and MRSA?

In the United States, approximately one-third of the population is colonized with Staph (89.4 million persons) and less than 1% is colonized with MRSA (2.3 million persons). Colonization rates with Staph are higher in the younger population, whereas colonization rates with MRSA are higher with the elderly.

Can a person get both Staph and MRSA infections at the same time?

Yes. However this is extremely rare. Most people are colonized and/or infected by one type of Staph. But it is not impossible to have infection with both types of Staph at the same time in different areas of the body.

Chapter II
How do we get Staph and/or MRSA infections?

Since approximately one-third of the population carry Staph bacteria somewhere on the skin, Staph infections are very common. Staph bacteria living on top of the skin can enter the body through a scratch or small wound. The object causing the wound is introduced to the outside layer of the skin (where the Staph lives). As the object pushes deeper into our skin, it carries with it the Staph bacterium which is then pushed deeper into the body tissues. Once it is deep inside, Staph can cause either minor or serious problems depending on a number of factors.

How are Staph and MRSA spread from one person to another?
Staph and MRSA bacteria are spread primarily through hand or skin-to-skin contact with someone who has these organisms on his or her skin, but can also be spread by touching objects that are contaminated with these bacteria. In other words, we touch a person or object that is

contaminated with Staph or MRSA, and then transfer this organism to our own body. How can this be prevented? *Simply by washing your hands!* Careful, scrupulous handwashing with soap and water, or waterless alcohol hand sanitizer, is the best defense against germs such as Staph and MRSA.

How long should you wash? If you are using soap and water, vigorously scrub your hands for 15-20 seconds, rinse well, then dry with a disposable towel, and use the towel to turn off the faucet (see also Appendix A).

What is the proper way to wash your hands using waterless alcohol hand sanitizer? Place a nickel-size amount of sanitizer in your hand. Rub the sanitizer thoroughly into your hands, paying particular attention to the thumbs and fingers, and then allow the product to air dry on your hands. Do <u>not</u> wipe the sanitizer off your hands with a towel, or your clothing, as air drying allows the product to kill the germs on your hands. *The simple act of handwashing, either by soap-and-water method, or using waterless alcohol hand sanitizer, is the single most important thing you can do to protect yourself against disease and to prevent any possibility of infection after contact with potential sources of contamination.*

MRSA and Staph infections have also been transmitted by athletic equipment (including equipment found in health clubs), tattoo equipment, razors, towels, and sheets that are shared by anyone who is colonized or infected with these organisms. Close-contact sports, such as wrestling and football, have also been contributing factors to the transmission of these organisms.

How did Staph become MRSA?
MRSA is a Staph bacterium that has become resistant to the antibiotic called methicillin. Actually, bacteria that have become resistant to antibiotics are on the rise throughout the world. A primary reason for this is the overuse of antibiotics. For years, physicians and hospitals have overprescribed antibiotics–using them unnecessarily and inappropriately. In addition, antibiotics are often

prescribed for viral illnesses, which are caused by viruses, not bacteria. Therefore, antibiotics cannot kill the viruses and this only contributes to antibiotic resistance. Patients demanding antibiotics for themselves or their children every time they experience illness have also contributed to this overuse. Surprisingly, however, the vast majority of antibiotics produced in the United States aren't used for human consumption but instead go into animal feed for cattle, chickens, and pigs – not to treat illness of the animal, but to accelerate growth of the animal and to prevent illness of those animals raised in overcrowded and unsanitary conditions. In addition, as antibiotics are excreted from these animals, the waste is oftentimes washed away into streams, rivers, and groundwater tables.

Another way that bacteria become resistant to antibiotics is through their ability to mutate. Once bacteria are introduced to antibiotics, they can, and often do, become resistant to them.

Can household pets transmit MRSA?

Yes. Recent evidence has shown that household pets, such as cats and dogs, can transmit MRSA to their owners. Perhaps the same might be true for Staph.

Who typically gets Staph or MRSA infections?

Infections with Staph and MRSA are more common and prevalent in hospitalized patients. These days, most patients admitted into hospitals are very sick, debilitated, immunocompromised, and even malnourished. These factors, coupled with the placement of devices such as indwelling urinary catheters, IVs, and other medical and surgical procedures, increase the likelihood of a patient acquiring an infection. However, Staph and MRSA infections are also prevalent in dialysis centers, skilled nursing facilities, and long-term care facilities where they manifest primarily as pneumonias, bloodstream infections, IV line infections, urinary tract infections, wound and bone infections.

Can healthy people get Staph and MRSA infections?

Yes. Staph and MRSA infections are also seen in otherwise healthy people who have either never been hospitalized, or have not been hospitalized for a long time. These infections manifest themselves as skin infections with initial lesions resembling those of a spider bite (round, raised, reddened, and painful).

What is "community-associated" or "community-acquired" MRSA (CA-MRSA)?

MRSA was never seen outside of hospitals or healthcare settings until 1999 when four previously healthy children who had never been hospitalized, suddenly died of overwhelming MRSA infections.

People with MRSA infections are said to have "community-associated" or "community-acquired" MRSA infections (or simply CA-MRSA) if they meet the following criteria:

They have _not_ had hospitalization or surgery, dialysis, residence in a long-term care facility, skilled nursing facility or hospice, during the past year; they have _no_ permanent indwelling catheters or percutaneous medical devices; they have _no_ medical history of MRSA infection and colonization; and they have the diagnosis of MRSA made in the outpatient setting or by culture positive for MRSA within 48 hours after admission to a hospital.

How common are CA-MRSA infections?

We have recently seen a marked increase in CA-MRSA infections in many parts of the USA, and in the world. The reason for this is not entirely clear, but it seems that the bacterium is now more prevalent and also more virulent (some strains have certain toxins that make them more invasive). MRSA can also be transmitted in the healthcare setting (called healthcare-associated MRSA or HA-MRSA), and it happens primarily through the contaminated gloves or hands of healthcare workers, or through contact among patients. Recent studies suggest that MRSA can also survive and replicate rapidly (1,000-

fold) inside a common type of amoeba (*Acanthamoeba polyphaga*) that is present on most surfaces in the hospital environment. Amoebas can also be spread in the air.

What are the risk factors associated with Staph and MRSA infections?

Factors associated with an increase in Staph and MRSA infections include poor hygiene, living in crowded places, and having skin abrasions or cuts, or being in contact with infected items. Studies have shown outbreaks of CA-MRSA in children, military recruits, Native Americans, Alaskan Natives, men who have sex with other men, and in people living in crowded environments and prisons.

Chapter III

What diseases can Staph and MRSA produce?

How is the infection produced?

Staph and MRSA infections are becoming more and more common in people who are otherwise healthy. Hygiene appears to play a role in whether or not somebody may get an invasive infection. The most common symptom of Staph or MRSA is the skin infection that starts as a boil or pimple. Many people blame it on a spider or insect bite. Interestingly, almost none of the people claiming such bites have seen the culprit. The common scenario is that they try to express the boil with their fingers. Sometimes pus is drained. But sometimes is not drained and more inflammation is produced in the skin, leading to a furuncle or large boil which subsequently gets larger, and can produce a localized cellulitis (redness and pain in the skin). Depending on whether or not the bacterium is a very aggressive one (i.e., if the bacterium infecting the skin produces some toxins), impetigo, furuncles, cellulitis, darkening, blistering, or necrosis of the skin is seen. Most people do not wash their hands or the wounds at the same time they try to drain the pimples. If the Staph is already colonizing the skin, even those boils produced by bacteria other than Staph will end up being infected with Staph due to the poor hygiene.

Why do some people with a Staph infection get very sick?

After infecting the skin, and depending on whether or not the person has a good immune response; the infection, after producing local symptoms, could proceed unchecked and enter the bloodstream producing bacteremia (growth of the bacteria in the bloodstream). Because the bacteria is not supposed to be in the bloodstream, certain cells in the body produce substances called "cytokines" which will produce fever and an inflammatory response. This inflammatory response is characterized clinically by fever, increased heart rate (tachycardia), increased respiratory rate (hyperventilation), and possible hypotension which may herald the starting of what is called "septic shock." Some patients manifest with deadly forms of overwhelming infection, or with the effects of Staph toxins which are called Scalded Skin Syndrome and Toxic Shock Syndrome.

What type of diseases can Staph and MRSA produce?
The diseases that Staph and MRSA can produce are very similar and they tend to be invasive. The diseases range from skin infections to pneumonia, infection of heart valves (endocarditis), infection of prosthetic devices (any device including pacemakers, hip or knee replacement devices, etc.), meningitis, spinal abscesses, and necrotizing fasciitis (necrosis of extensive areas of skin that can quickly become fatal). Staph and MRSA infections can also be seen after surgery, or after indwelling devices are placed (including lines for IV antibiotics or parenteral nutrition, shunts placed in the brain for hydrocephalus, etc.). The mortality rate among patients with MRSA infections appears to be substantially higher than among those infected with Staph (MSSA).

Is MRSA more aggressive than Staph?

Yes, but we used to think differently. However CA-MRSA is more dangerous and virulent than Staph (MSSA) and is more difficult to treat. The current epidemic strains of CA-MRSA have the ability to produce certain toxins that can deactivate white blood cells (Panton-Valentine leukocidin toxin for example), leading to severe skin disease (necrotizing fasciitis) or necrotizing pneumonia. This toxin is less often found in HA-MRSA. CA-MRSA tends to be more susceptible than HA-MRSA to antibiotics, with the exceptions of beta-lactams and erythromycin.

Chapter IV
How do we diagnose a Staph or MRSA infection?

Staph or MRSA infections are revealed by collecting samples in a sterile environment, and sending them to the microbiology laboratory for culture. MRSA is revealed by showing that the Staph in culture is resistant to oxacillin in the petri dish. The process consists of two steps. First the bacterium needs to be cultured from the sample. Once it is isolated, it is exposed to the antibiotics to determine if there is resistance to any of them. Some new molecular tests are used to speed up the diagnosis. The sample should be taken aseptically from skin lesions, bloodstream, urine, and sputum. It can also be taken from sterile body fluids (cerebrospinal fluid, pleural fluid, joint fluid). In general, any body tissue or fluid can be cultured. Lines and indwelling devices can also be cultured. Additionally, after surgery, if a prosthetic material is thought to be infected, it can be sent for culture.

How do we diagnose Staph or MRSA colonization?

Colonization with Staph or MRSA is typically diagnosed by wiping both nostrils with a swab and sending the swabs for culture to the laboratory. Many people have the tendency to touch their nose, and children do it very frequently. Since the Staph or MRSA inhabits the skin, the fingers carry bacteria that subsequently remain in the moist environment of the nose.

How do we decolonize a person?

If the nose is colonized, an ointment (mupirocin) is called for. Additionally, oral antibiotics are taken for a period of at least 7 days. Subsequently, cultures are repeated to determine whether or not the bacterium has been eradicated. This eradication may be transient since re-infection can occur as a result of manipulation of the nose with infected/colonized fingers. Infections with S. *aureus* and MRSA are more frequent in people who are already colonized. Therefore, it is important to try to reduce the risk of infection by pursuing decolonization.

Chapter V
How do we treat Staph and MRSA infections?

Staph and MRSA infections are treated with antibiotics. If there is any collection of purulent material (pus) in the skin or in an internal organ or cavity it needs to be drained.

Antibiotics are not able to dissolve the pus that exists already in a large abscess. Therefore, such an abscess needs to be surgically drained, or drained with a needle using the help of radiographic techniques such as computer tomography or ultrasound. In the case of infections which are not severe, a short course of antibiotics might suffice. However, when infections are deep or severe, antibiotics, preferably delivered intravenously, are needed for several weeks.

Infections with Staph and MRSA need to be taken seriously, as they have the potential to be lethal within a short period of time if not treated promptly. The strains responsible for CA-MRSA infections seem to have an increased virulence; and in matter of days, infections with CA-MRSA can produce large abscesses.

Among the antibiotics useful in the treatment of MRSA, vancomycin, an intravenous antibiotic, has been traditionally used. Resistance to vancomycin is extremely rare, and only a few cases have been reported. Other antibiotics that can be used to treat MRSA include linezolid, daptomycin, tigecycline, clindamycin, trimethoprim-sulfamethoxazole, doxycycline, rifampin (not recommended alone), and quinupristin-dalfopristin.

Chapter VI
How do we prevent infections with Staph and MRSA?

The cornerstone for prevention of infections with Staph and MRSA is the implementation of the simple and straightforward measures indicated below:

Wash your hands with soap and water regularly, in particular after using the toilet and before meals. An antiseptic soap can be used, but there is not much data to support the idea that it is better than plain soap and water. You can also use a hand sanitizer with alcohol, if soap and water are not available. The goal is to maintain clean hands. Good hygiene should be maintained along with daily bathing.

If you injure your skin by cutting yourself or scraping the skin enough to bleed, clean your wound as soon as possible. Then cover it with a bandage until the skin is intact again. If redness appears or pain increases, seek medical help.

Don't touch other people's broken skin, boils, infected or bandaged areas. If you must, wash your hands before and after touching their skin.

Don't share towels, clothing, bar soaps, shampoos, nail clippers, cosmetics, razors or toothbrushes.

If you play competitive sports, do not share equipment that has not been cleaned between users. Shower immediately after competing.

Discourage allowing pets to share beds with members of the household, as there is evidence that household pets might transmit MRSA. If your pets play in the dirt or are outdoors frequently, the idea of sharing beds or sofas with them is probably not a good one. If you do so, make sure your pets are bathed properly.

Try to avoid walking on the floor in your bare feet, and if you do so, wash them regularly.

If you ever notice a place on your body that looks like an insect bite, spider bite, or a persistent sore, do not ignore it. See your healthcare provider immediately if severe itching or pain develops. Don't wait for blistering before you seek treatment.

If you are already colonized, follow the instructions of your healthcare provider that may include applying a mupirocin ointment to the inside of your nose, taking oral antibiotics, and practicing good hygiene. Additionally, using shampoos and antiseptic soaps containing chlorhexidine has been suggested. For specific measures, follow the advice of your healthcare provider. These measures should help to prevent the number of recurrences of infection. Note that resistance of MRSA to mupirocin has been documented. An oral antibiotic should be used in conjunction with the mupirocin ointment to help in the eradication of MRSA from the nostrils.

If you have a history of recurrent infections with Staph or MRSA, and have a sudden unexplained onset of chills or high fever, consider contacting your healthcare provider as this might herald an underlying infection.

How can we prevent transmission of Staph and MRSA in the healthcare setting?

In the healthcare setting, prevention of transmission from patient to patient is achieved through the use of Standard Precautions and Contact Isolation. The goal is to prevent infection by direct contact and droplet spread.

In the healthcare setting, the patient with MRSA infection (and MRSA colonization!) is placed on Contact Isolation, preferably in a private room; or if not possible, in a room with another patient with MRSA, but no other infection. A patient is placed on Contact Isolation before demonstration by cultures if there is a hint or suspicion of MRSA infection. Gloves and gowns are worn when entering the room, and gloves are worn when handling all devices and laundry. Masks are used if the patient has pneumonia, or if there is a possibility of transmission by droplets (e.g., changing dressings in abdominal wounds, or if he/she is sneezing). Handwashing or use of sanitizer is mandatory before entering and after leaving the room. Patients should have in their rooms stethoscopes and other material that should not be shared with other patients. Dishes and dietary trays are <u>not</u> a major source of disease transmission.

Contaminated dressings and devices need to be disposed of properly in specific containers. Contaminated equipment and surfaces should be cleaned with a solution of 1:100 bleach and water, or a commercial disinfectant. Linens should be washed in hot water and dried with a high-temperature dryer setting.

Is surveillance culturing and isolation of patients useful?
Active surveillance culturing to identify MRSA-colonized patients, and the use of isolation precautions have been advocated and have been used in the healthcare setting successfully. These measures should help to reduce the spread of the multidrug-resistant strains of MRSA seen in

hospitals. Although it is well known that antibiotic usage exerts pressure to select for resistant flora, patients acquire MRSA in the hospitals primarily due to contamination. This underscores the importance of contact precautions and active surveillance cultures.

Chapter VII
Patient Vignette 1

Mark is a 35-year-old single male who lives in the suburbs of Chicago. He works in a plastic factory where he handles boxes which are loaded with the finished products that are exported or sent to other locations in the US. He uses gloves at work and does not smoke, drink or use any drugs. He has a dog that is three years old. His dog sleeps with him, and Mark has noticed some lesions on his dog's skin. He is aware that his dog has skin allergies and that his dog sometimes gets severe itching due to flea allergies. He uses medication to treat the dog's condition every three months.

Mark has noticed lately that he himself has small lesions on his arms and shins that appear to come from nowhere, and he develops localized swelling whenever he scratches them due to itching. He thought for some time that these were allergic reactions to his dog's dander, and has increased the bathing of his dog to once a week. However, he keeps having the small lesions from time to time. They improve after few days, but there are always small scabs that seem to form one or two days after he scratches them. He has noticed that some of them itch and appear like small blisters, similar to spider bites, that subsequently enlarge. However, he has not seen any spiders in his house or his bed.

It was Friday afternoon and Mark wanted to go out with friends to a local restaurant to have dinner. He noticed that he had a small pimple on his right leg that had appeared that morning. He had carefully washed it out and had applied some over-the-counter antibacterial ointment on it. He noticed that there was a small area of swelling around the lesion. He also experienced some increased itching. He took a shower and went to dinner. He really had a good time, drank two beers and went home.

The next morning, he awoke with extreme pain in his right leg and he felt feverish. He looked at his leg and saw that there was a large red and swollen area. He touched his leg and felt severe pain. He changed his clothes and went to the hospital where he was seen in the Emergency Department. He was evaluated by a physician and had some blood work done. Then he was informed that he needed to remain in the hospital for intravenous antibiotics because he had a high fever and his blood work showed an increase in white blood cells. A technician took some blood samples as well as blood cultures from him. He was asked if he was allergic to any antibiotics. He had no allergy to any antibiotic, and was given an intravenous antibiotic called cefazolin. A nurse came and did cultures of his wound which was then draining pus and was darker. His leg did not seem to improve with the antibiotic, and his leg was getting more swollen. Thereafter, another doctor came and told him that his antibiotic would be changed to a more powerful one called vancomycin. The pain in his leg was not getting better with the painkillers, but nevertheless, he tried to sleep. He had an MRI done on his leg to see if the infection had rapidly progressed and affected his bones. The next day, the redness and swelling in his leg were a little better. The pain, however, was still there and was severe. That afternoon, he was told by hospital staff that people coming into his room would need to wear gloves and gowns. He got worried, thinking that he might have some highly contagious disease. He was told that it was the hospital's policy, that the infection was rather commonly seen, and that its name was MRSA.

The doctor that saw him the day before told him that he needed to culture his nose to see if the MRSA was hiding in his nostrils. He took swabs of the inside of Mark's nostrils. During the procedure, Mark was trying hard not to sneeze and started to have profuse tearing. With the new antibiotic, his leg was rapidly improving and the swelling was going down. The MRI of his leg came back negative for bone infection, and he was told that he needed either intravenous or oral antibiotics at home for at least seven more days.

The next day, Mark's leg was almost back to normal except for the dark scab and the ulcer. He was told that his nose culture was positive for MRSA, and that he needed to apply an ointment to his nostrils with a cotton tip three times a day for at least seven days. The name of the ointment was mupirocin, and he was also told that he needed to take two pills a day for seven days. Their names were rifampin and trimethoprim-sulfamethoxazole. One of them would make his urine dark orange in color. His blood cultures came back negative, and no further workup was planned. He was getting anxious as he needed to go back to work. Since his leg was looking much better, he was finally discharged and told to take another pill called linezolid for seven more days, and to follow up with his primary care physician. He was also told to try to shower and shampoo his hair with a chlorhexidine gluconate solution.

Mark's case is a very common one these days. It is just one of the hundreds of cases of CA-MRSA infection seen daily in emergency departments. Mark was able to seek prompt care, was treated adequately, and he recovered successfully.

Patient Vignette 2

Jeanne is 70 years old and lives in New York in an assisted-living community. She has diabetes mellitus and heart disease. She had a coronary bypass graft six years ago. She has also had two knee replacements. She had a right hip replacement after a fracture following a fall the year before her heart surgery. She still has some problems ambulating and has some pain now and then in her hip. Her blood sugar level has been well controlled, and in the mornings it is in the 105 to 110 range. She has no kidney disease or diabetic neuropathy although she suffers from glaucoma and has had visual troubles.

It is late December and she has gotten a flu and a pneumonia shots two months earlier. She has been doing well, despite the fact that some of the people in her assisted-living community were complaining of colds, and two residents have recently had pneumonia, for which they needed to be hospitalized.

Jeanne had finished lunch and went to take a nap in her room. She awoke about three hours later with chills and nausea. She also felt dizzy, so she called the personnel in her assisted-living facility. An aide came to her room and took her temperature. She had a fever of 102 degrees. After calling her doctor, the decision was made to send her to the Emergency Department of the nearby hospital. She agreed. She was feeling more and more ill. She was short of breath and beginning to cough.

After she was transported to the Emergency Department, she was evaluated by a nurse and then by one of the physicians on duty. She was found to have a temperature of 104 degrees. Her white blood cell count was elevated and her chest x-ray showed a right-side pneumonia. The decision was then made to admit her to the hospital. Blood work was done, including cultures of her blood and her urine.

Jeanne was beginning to cough more and more, she was short of breath and had some chest tightness. She had an electrocardiogram done that showed some abnormality suggesting the possibility of a heart attack. Her oxygenation was also poor, which led to her having oxygen delivered to her by a mask. She was soon wheezing constantly. She was given a diuretic and some bronchodilator nebulizations. After few minutes, she began to feel better. She also received two intravenous antibiotics.

Because of the various medical problems she was facing. The decision was made to have her admitted to the Intensive Care Unit. Her breathing was still labored two hours later. A cardiologist on call saw her and recommended more diuretics, and also an Infectious Diseases consultation following his observation that she had high fever and her blood pressure was going down.

Her low blood pressure prompted the use of some medications to increase it. Also, a catheter was placed by the cardiologist in her neck to better evaluate her cardiac condition. After a few minutes, the cardiologist concluded that Jeanne was having an early case of sepsis.

The Infectious Diseases specialist arrived one hour later and recommended that more antibiotics be started, and requested a CT scan of Jeanne's chest and for an echocardiogram to be done in the morning. Jeanne was feeling increasingly weak, and her breathing remained heavy despite all the treatments.

Three hours later, she was on an oxygen mask and still unable to breathe properly. Additionally, her urine was starting to diminish (she had a urine catheter placed earlier to measure her urine). The decision was then made by the attending physician to put her in a mechanical respirator, and she agreed. She was then fully sedated prior to her placement in the mechanical respirator.

The next morning, her sedation was reduced slightly to see if she was able to understand and follow commands. Her CT scan showed that she had pneumonia in her right lung.

Her echocardiogram did not show any sign of infection on the heart valves, but her blood cultures came back positive for possible Staph infection. Twenty-four hours more were needed to have the cultures finalized, and to ascertain whether or not she had a Staph infection. Her blood pressure remained low, and she was on three medications in an effort to increase it. Her urine output had become very minimal, and a consultation was requested with the renal consultant (nephrologist).

Jeanne was on the ventilator, and samples of her lung secretions (sputum) were sent for culture to see what type of bacteria may have caused her pneumonia. Her urine cultures were negative for bacteria, and she was started on tube feedings using a tube going from her mouth to her stomach. The culture of her sputum demonstrated the presence of MRSA as well.

Over the next few days her clinical condition did not change much. Her blood cultures turned out to be positive for MRSA. She continued on the intravenous antibiotics, and subsequent blood cultures were negative, suggesting that the antibiotics she was getting were effective against MRSA. However, she continued to retain fluid and had no urine output. The renal consultant recommended dialysis until her kidneys recovered. She agreed to the placement of a catheter in her left groin to enable the dialysis. The first session of dialysis proceeded well, and her apprehension about it was reduced. She was still intubated and anxious since she could not talk, although the nurses and doctors could understand what she was trying to convey by nodding her head. Her frustration came from the fact that she could not say much except to nod "yes" or "no" when she was asked questions.

One week later, and after three sessions of dialysis, her clinical condition clearly started to improve. She began to have a scant amount of urine, signaling that her kidneys were improving. Her chest x-rays, however, still showed pneumonia, and some fluid started to accumulate in her right pleural cavity (the space between the lungs and the chest wall). She became feverish again, and the blood, urine, and sputum cultures were repeated. A procedure was performed to have the fluid removed using a needle. The fluid turned out to be purulent, suggesting infection, probably with MRSA. Her condition was called "empyema."

She had a chest tube inserted, and she was told that she would probably need a procedure later to clean her pleural cavity once her condition further improved. Her oxygen needs diminished after the fluid was removed, and her fever was reduced. The blood cultures and urine tests that were done did not show any bacteria. MRSA was again found on the culture of her sputum.

Eventually, she was slowly weaned off mechanical ventilation and then released from the respirator. She still had the chest tube in place, but her breathing was much better, and the pneumonia was clearly improving, according to the chest x-rays. She developed back pain at the site of the chest tube insertion. Two days after her release from the respirator she was able to have a clear liquid diet. The plan was to have a procedure called "decortication" to cleanse her pleural cavity of the remaining purulent material once she was sufficiently recovered.

Jeanne was feeling stronger the following week, and her decortication was done by the middle of the week. The procedure went without complications, but she was in severe pain afterward. An intravenous pump for self-administration of morphine was set up, and finally, she had more control over her back pain. She still had a chest tube, and the plan was to remove it once the amount of fluid that was draining diminished.

Five days later, the chest tube was removed, and Jeanne was eating solid food. The analgesic pump was removed, and she was able to sit down in her chair and spend a few hours there daily. Since she had improved significantly, she was transferred to another floor. She was seen by the discharge planner for her case and was told that she would go to a rehabilitation center for a while before returning to her assisted-living complex. Her urinary catheter was removed, and as she was able to go to the commode by herself.

The physical therapist worked with her on a daily basis. Initially, Jeanne was not able to walk much. In fact, she was not even able to stand up without having her legs give out. Little by little, she improved and gained the confidence to walk the ward floor several times a day, at first with a walker and then by herself. Her intravenous antibiotics were continued through a long-term intravenous line placed for that purpose. She finally was encouraged to go to a rehabilitation center, and she arranged for one of her relatives to visit the facilities prior to choosing one.

The following week, she was transferred to the rehabilitation center where she spent two weeks finishing her intravenous antibiotics while making new friends and doing an increased amount of physical therapy. She went to see her primary care physician and then the Infectious

Diseases specialist the next day. The Infectious Diseases specialist reviewed her chest x-rays and laboratory workup, and told her that he was pleased with her improvement. He discontinued her intravenous antibiotics and removed the IV line that she had been using for antibiotic administration. She was to follow up with him two weeks later and have a repeat chest x-ray and some laboratory work done, which included a repeat of blood cultures.

Jeanne was told that she could return to her assisted-living community. The next day, she returned to her home and was greeted by her neighbors and given a small welcome-back party. Her laboratory workup and blood cultures were negative. In the following months after her illness, Jeanne did not have a recurrence of her infection.

Jeanne's case illustrates the potential of MRSA to produce severe disease. In the end, Jeanne's case turned out well. However, sometimes complications arise with elderly patients due to the relatively compromised state of their immune systems. Many of those patients have several pre-existing chronic or debilitating diseases. Some of them suffer from evident malnutrition due to feeding problems, inability to swallow properly, or simply because they do not have much appetite. Those patients have a poor prognosis if they also have an acute infection with Staph or MRSA.

Chapter VIII
Conclusions

Infections with Staph and MRSA are a source of significant morbidity and mortality. In certain areas, the number of cases of persons having skin infections with MRSA is reaching epidemic proportions. It is a public health threat that must not be ignored because it is potentially lethal. To effectively fight this infection and to reduce its spread, requires a significant behavioral change. These "community-associated" MRSA infections, unlike the old Staph infections that caused less invasive infections, spread extremely rapidly, and in a matter of hours, can produce significant tissue damage leading to necrotizing lesions. Good hygiene and increased public awareness will help to stop the spread of CA-MRSA. The best way of reducing the spread of MRSA, in addition to treating the infections, is to prevent people from getting it.

Photographs

The photographs that follow were taken from patients who gave their full consent before the pictures were obtained. These photographs represent what is frequently seen in hospitals and clinics across the United States as well as in other parts of the world.

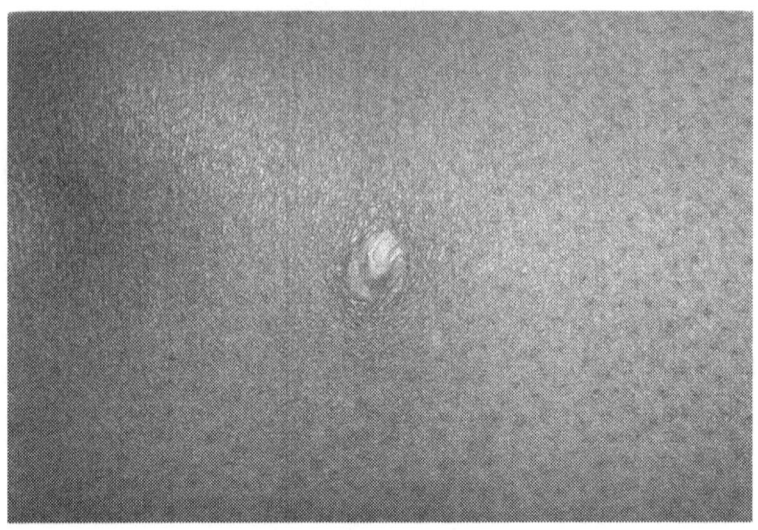

Figure 1. Patient with blistering lesion in her leg. Culture revealed MRSA.

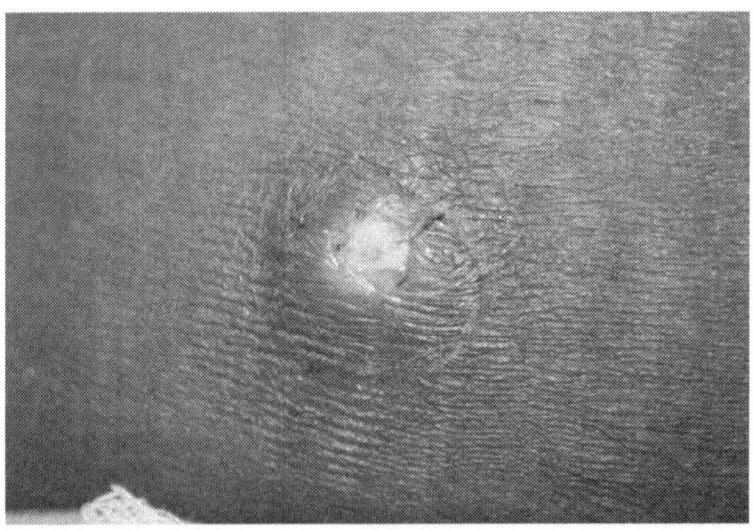

Figure 2. Patient shown in Figure 1 with abdominal ulcer.

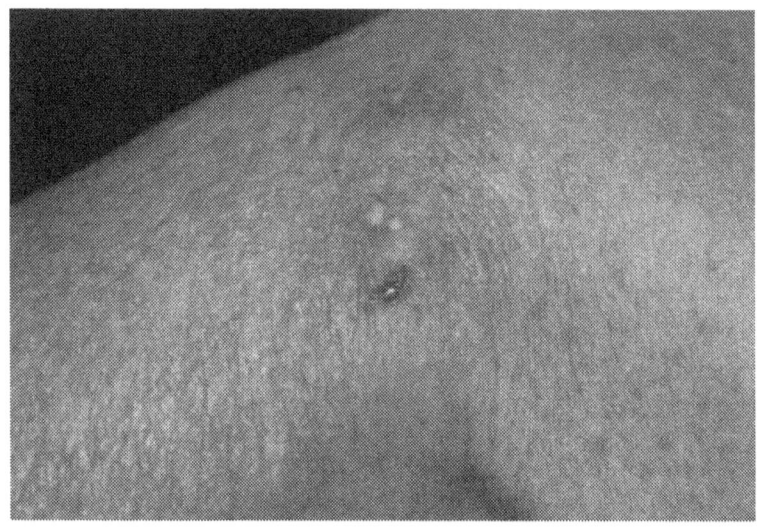

Figure 3. Patient who had four days earlier a "spider bite-like"lesion. Started on antibiotics. Culture revealed Staph.

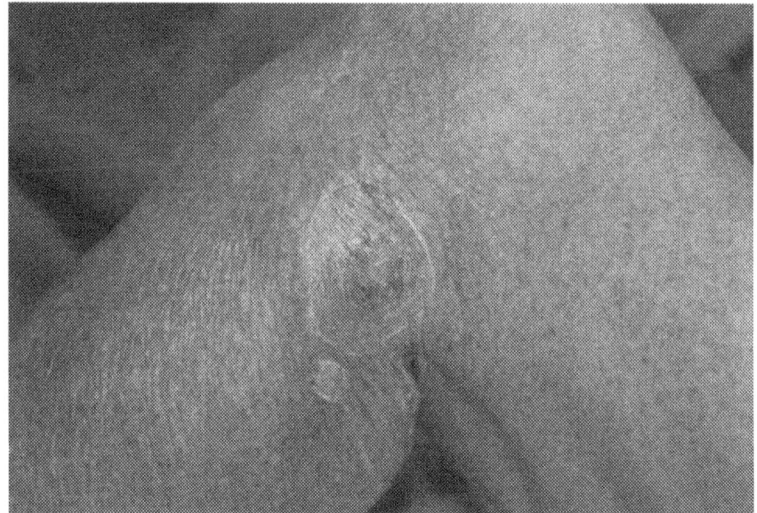

Figure 4. Patient shown in Figure 3, after 48 hours.

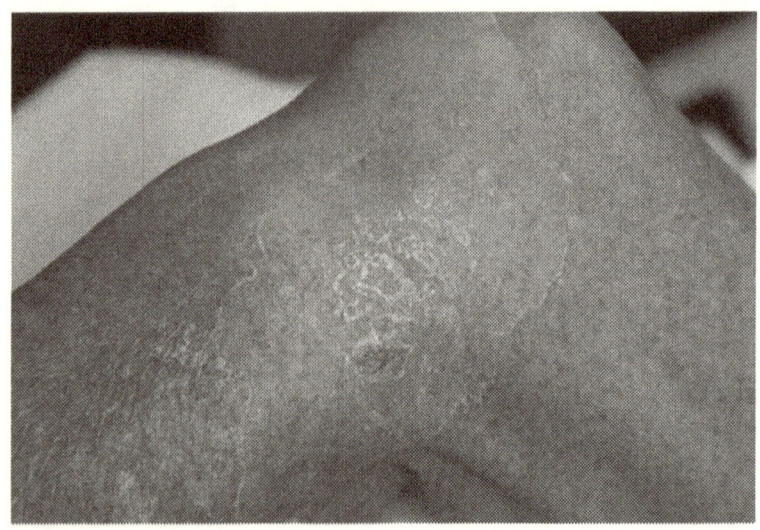

Figure 5. Patient shown in Figure 3, seven days later.

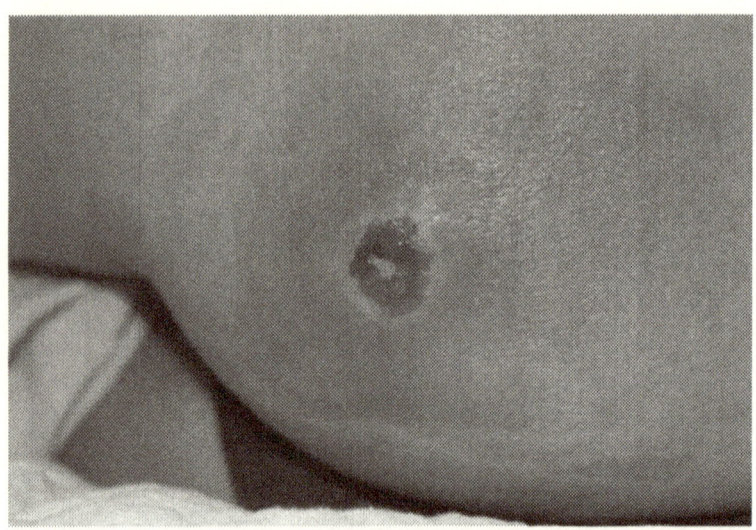

Figure 6. Patient with MRSA buttock ulcer and severe pain four days after a "spider bite-like" lesion.

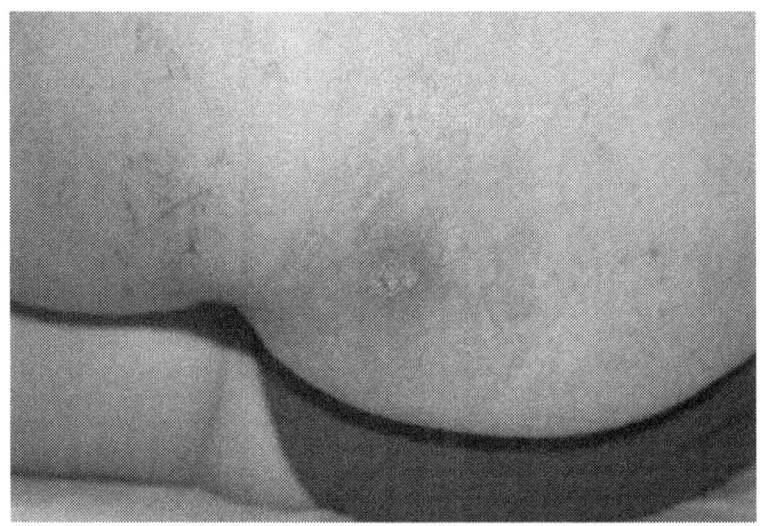

Figure 7. Patient shown in Figure 6. After ten days of antibiotics.

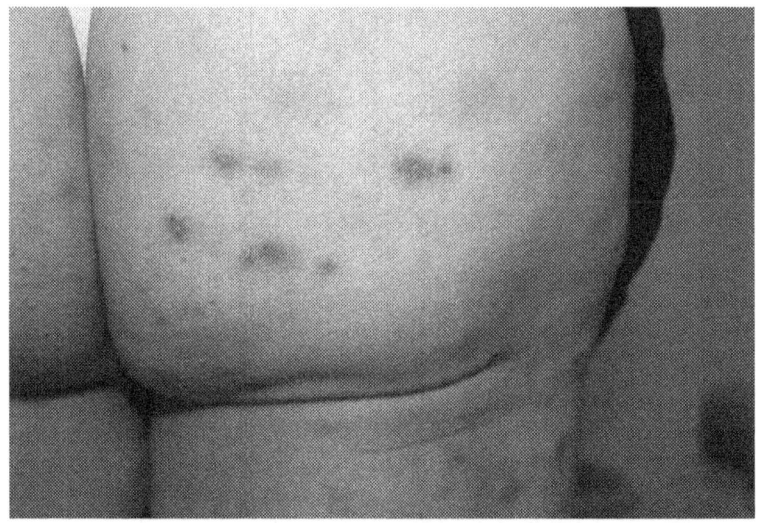

Figure 8. Patient with history of recurrent MRSA skin infections over a period of two years.

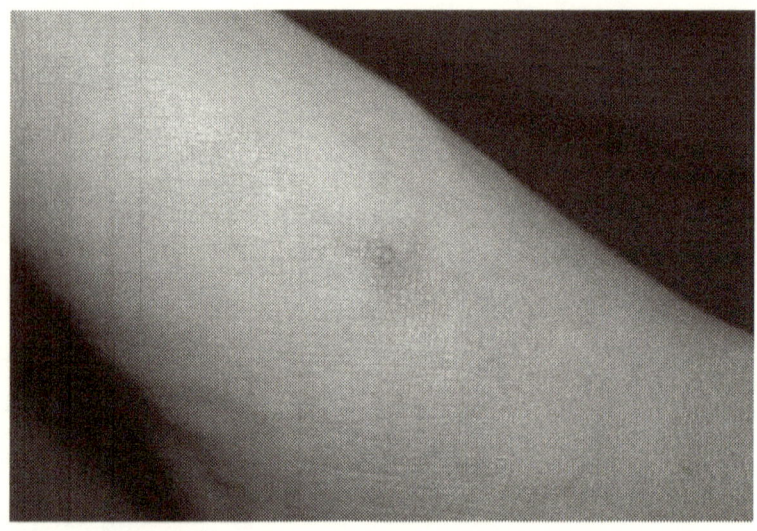

Figure 9. Patient with recurrent "spider bite-like" lesions, with itching, redness, and blistering.

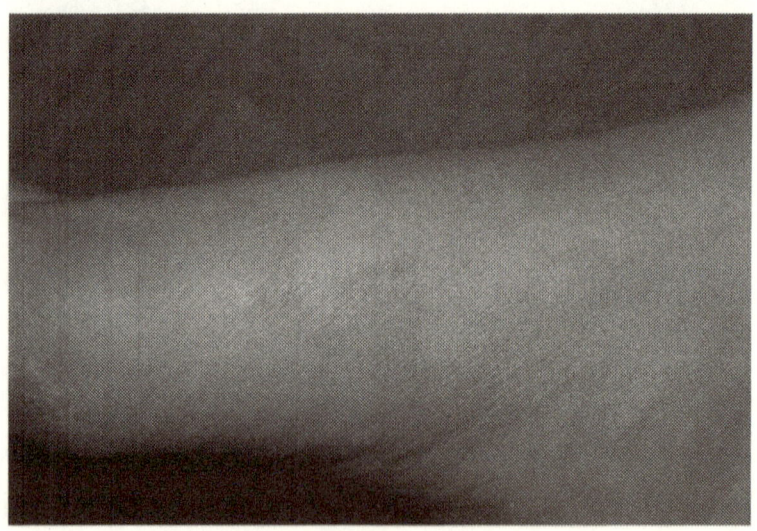

Figure 10. Same lesion 24 hours after using mupirocin.

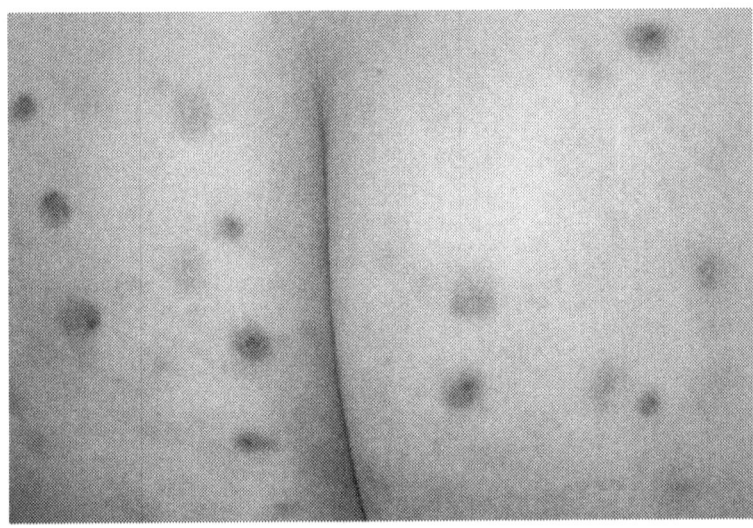

Figure 11. Patient after several weeks of itching.

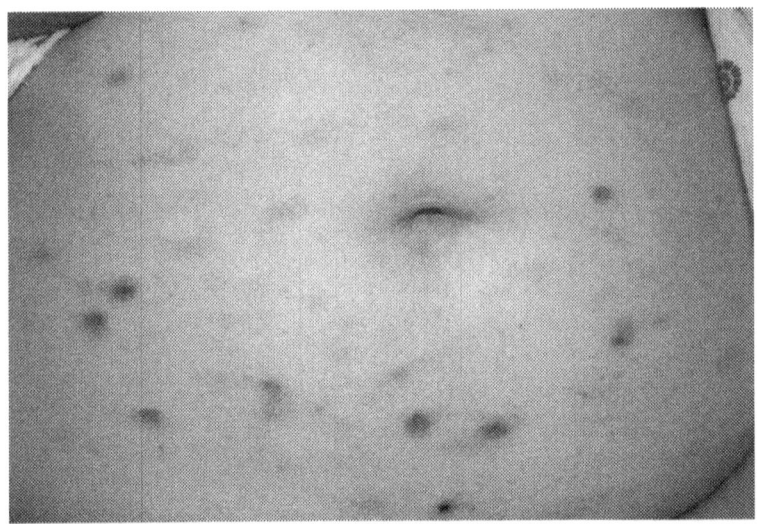

Figure 12. Patient shown in Figure 11 with multiple abdominal lesions. Cultures showed Staph.

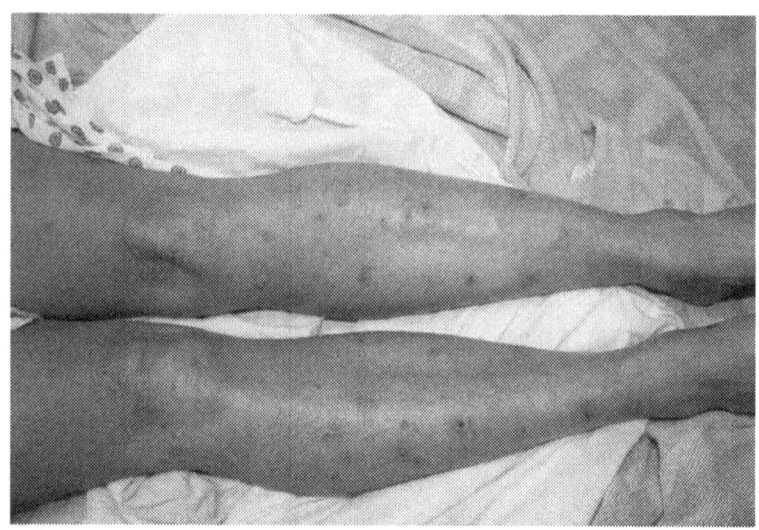

Figure 13. Legs of patient shown in Figure 11.

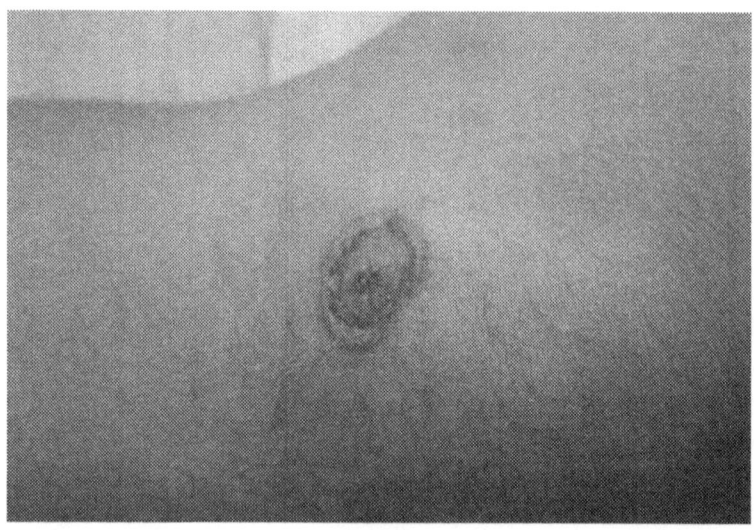

Figure 14. Another patient with a lesion three weeks old. Culture revealed Staph.

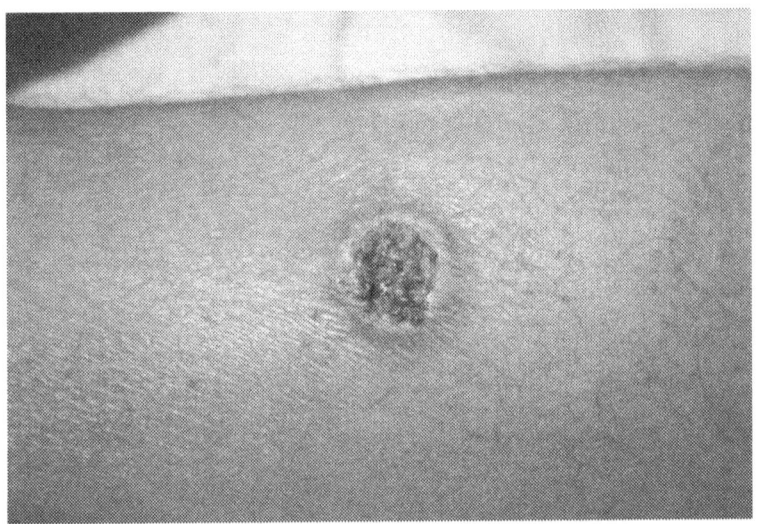

Figure 15. Patient shown in Figure 14 with necrotic lesion.

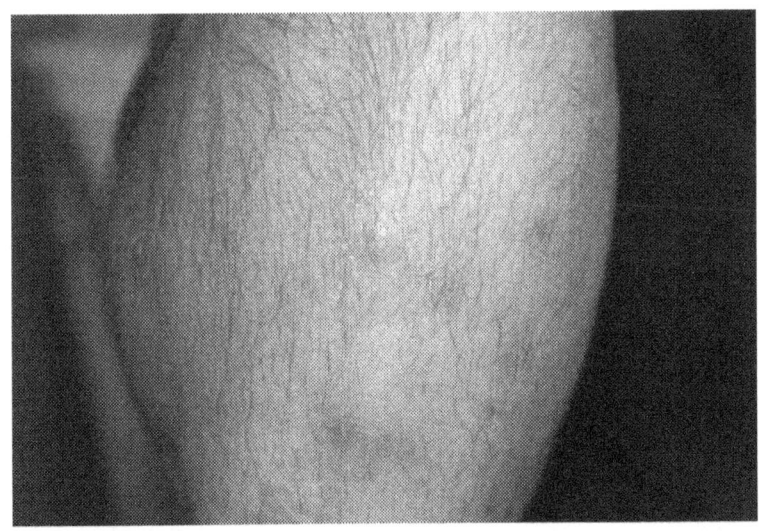

Figure 16. Right forearm of a patient with history of recurrent MRSA infections.

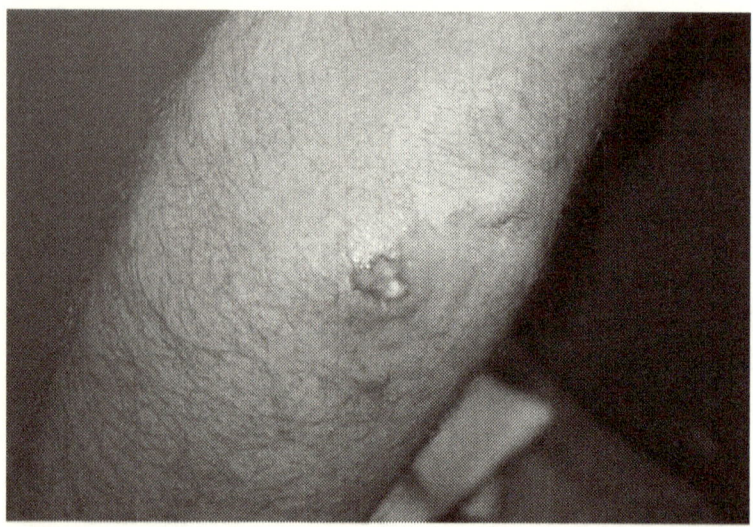

Figure 17. Left elbow of patient shown in Figure 16. The wound has been draining for several days.

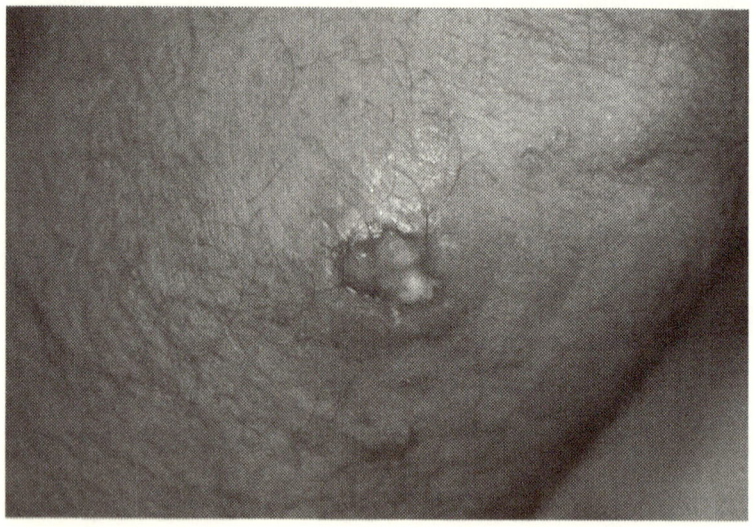

Figure 18. Closer view of elbow showing ulcer.

Appendix A
Handwashing

It might appear redundant, but unfortunately most people do not wash their hands correctly. It is preferable to use gloves if you are going to touch a skin lesion or a device that might be contaminated with Staph or MRSA. Do the following before and after touching potentially contaminated surfaces.

If you use soap and water:

Apply liquid soap or use clean bar of soap after putting your hands under running lukewarm water.

Rub your hands for at least 15-20 seconds, taking care to clean the palms, under the fingernails, and the backs of the hands.

Rinse your hands well with water.

Use a clean paper towel to dry your hands. Use the paper towel to turn off the faucet so as to avoid contaminating your hands again!

If you use a sanitizer, it is preferable to use only the alcohol-based ones.

Appendix B
Mupirocin for Nasal Decolonization

Mupirocin is a topical antibacterial sold at 2% concentration in cream and ointment forms. The ointment is better than the cream. Before using this medicine you need to contact your healthcare provider, especially if you are pregnant, have multiple allergies, or if you take other medicines.

To apply the ointment:

Wash your hands, blow your nose, and wash your hands again.

Put a small amount (¼ of an inch) of mupirocin ointment on your clean little finger or a cotton bud (if you are applying it to another person use only the cotton bud method, don't use your fingers!) and apply it all around inside of the nostril. Repeat with the other nostril. This needs to be done several times a day (3-4 times) as the ointment melts down. The ointment can cause sneezing, or a stinging sensation. If the stinging sensation does not disappear, stop using the ointment and contact your healthcare provider.

Wash your hands after application. This ointment needs to be used for approximately 5-7 days consecutively, together with some oral antibiotics prescribed by your healthcare provider.

Throw away any ointment left over after finishing the treatment.

Keep the ointment away from children. If you get the ointment in your eyes, rinse with water and call your healthcare provider immediately.

Note that effective decolonization of the nose entails the use of mupirocin and oral antibiotics. Mupirocin alone will probably not be sufficient to successfully eradicate Staph or MRSA from the nostrils.

Appendix C
Chlorhexidine for Skin Decolonization

In patients that have skin colonization (carriage), chlorhexidine gluconate has been used successfully to eradicate MRSA. Showering twice a day is recommended in order to reduce the skin's bacterial load.

Use commercially available chlorhexidine gluconate 4% solution instead of soap during showering. It can be used also as a shampoo. Rinse well and dry skin. Discontinue use immediately if redness or skin allergy appear, and seek medical advice.

Bibliography

1. Adem P.V., Montgomery C.P., Husain A.N., Koogler T.K., Arangelovich V., Humilier M., Boyle-Vavra S., Daum R.S. *Staphylococcus aureus* sepsis and the Waterhouse-Friderichsen syndrome in children. **N. Engl. J. Med. 2005;353:1245-1251.**

2. Benner E.J., Kayser F.H. Growing clinical significance of methicillin-resistant *Staphylococcus aureus.* **Lancet 1968;2:741-744.**

3. Bischoff W.E., Wallis M.L., Tucker B.K., Reboussin B.A., Pfaller M.A., Hayden F.G., Sherertz R.J. "Gesundheit!" Sneezing, common colds, allergies, and *Staphylococcus aureus* dispersion. **J. Infect. Dis. 2006;194:1119-1126.**

4. Bootsma M.J., Diekmann O., Bonten M.J. Controlling methicillin-resistant *Staphylococcus aureus*: quantifying the effects of interventions and rapid diagnostic testing. **Proc. Natl. Acad. Sci. USA 2006;103:5620-5625.**

5. Coia J.E., Duckworth G.J., Edwards D.I., Farrington M., Fry C., Humphreys H., Mallagan C., Tucker D.R., Joint Working Party of the British Society of Antimicrobial Chemotherapy, Hospital Infection Society, Infection Control Nurses Association. Guidelines for the control and prevention of meticillin-resistant Staphylococcus aureus (MRSA) in healthcare facilities. **J. Hosp. Infect. 2006;63 (Suppl. 1):S1-S44.**

6. Eriksen N.H., Espersen F., Rosdahl V.T., Jensen K. Carriage of *Staphylococcus aureus* among 104 healthy persons during a 19-month period. **Epidemiol. Infect. 1995;115:51-60.**

7. Francis J.S., Doherty M.S., Lopatin U., Johnston C.P., Sinha G., Ross T., Cai M., Hansel N.N., Perl T., Ticehurst J.R., Carroll K., Thomas D.L., Nuermberger E., Bartlett J.G. Severe community-onset pneumonia in healthy adults caused by methicillin-resistant *Staphylococcus aureus* carrying the Panton-Valentine leukocidin genes. **Clin. Infect. Dis. 2005;40:100-107.**

8. Gillet Y., Issartel B., Vanhems P., Fournet J.C., Lina G., Bes M., Vandenesch F., Piemont Y., Brousse N., Floret D., Etienne J. Association between *Staphylococcus aureus* strains carrying gene for Panton-Valentine leukocidin and highly lethal necrotizing pneumonia in young immunocompetent patients. **Lancet 2002;359:753-759.**

9. Graham P.L. 3rd, Lin S.X., Larson E.L. US population-based survey of *Staphylococcus aureus* colonization. **Ann. Intern Med. 2006;144:318-325.**

10. Grundmann H., Aires-de-Souza M., Boyce J., Tiemersma E. Emergence and resurgence of methicillin-resistant *Staphyloccoccus aureus* as a public-health threat. **Lancet 2006;368:874-885.**

11. Huskins W.C., Goldmann D.A. Controlling methicillin-resistant Staphylococcus aureus, aka "Superbug." **Lancet 2005;365:273-275.**

12. Huws S.A., Smith A.W., Enright M.C., Wood P.J., Brown M.R. Amoebae promote persistence of epidemic strains of MRSA. **Environ. Microbiol. 2006;8:1130-1133.**

13. Kazakova S.V., Hageman J.C., Matava M., Srinivasan A., Phelan L., Garfinkel B., Boo T., McAllister S., Anderson J., Jensen B., Dodson D., Lonsway D., McDougal L.K., Arduino M., Fraser V.J., Killgore G., Tenover F.C., Cody S., Jernigan D.B. A clone of methicillin-resistant *Staphylococcus aureus* among professional football players. **N. Engl. J. Med. 2005;352:468-475.**

14. Kluytmans J., van Blekum A., Verbrugh H. Nasal carriage of *Staphylococcus aureus*: epidemiology,

underlying mechanisms, and associated risks. **Clin. Microbiol. Rev. 1997;10:505-520.**

15. Kuehnert M.J., Kruszon-Moran D., Hill H.A., McQuillan G., McAllister S.K., Fosheim G., McDougal L.K., Chaitram J., Jensen B., Fridkin S.K., Killgore G., and F. C. Tenover. Prevalence of *Staphylococcus aureus* nasal colonization in the United States, 2001-2002. **J. Infect. Dis. 2006;193:172-179.**

16. Lina G., Piemont Y., Godail-Gamot F., Bes M., Peter M.O., Gauduchon V., Vandenesch F., Etienne J. Involvement of Panton-Valentine leukocidin-producing *Staphylococcus aureus* in primary skin infections and pneumonia. **Clin. Infect. Dis. 1999;29:1129-1132.**

17. Lowy F.D. *Staphylococcus aureus* infections. **N. Engl. J. Med 1998;339:520-532.**

18. Miller L.G., Perdreau-Remington F., Rieg G., Mehdi S., Perlroth J., Bayer A.S., Tang A.W., Phung T.O.,

Spellberg B. Necrotizing fasciitis caused by community-associated methicillin-resistant *Staphylococcus aureus* in Los Angeles. **N. Engl. J. Med. 2005;352:1445-1453.**

19. Miller M.A., Dascal A., Portnoy J., Mendelson J. Development of mupirocin resistance among methicillin-resistant *Staphylococccus aureus* after widespread use of nasal mupirocin ointment. **Infect. Control Hosp. Epidemiol. 1996;17:811-813.**

20. Moran G.J., Krishnadasan A., Gorwitz R.J., Fosheim G.E., McDougal L.D., Carey R.B., Talan D.A., for the EMERGEncy ID NET Study Group. Methicillin-resistant *S. aureus* Infections among patients in the Emergency Department. **N. Engl. J. Med. 2006;355:666-674.**

21. Muto C.A., Jernigan J.A., Ostrowsky B.E., Richet H.M., Jarvis W.R., Boyce J.M., Farr B.M. SHEA guideline for preventing nosocomial transmission of multidrug-resistant strains of *Staphylococcus aureus* and enterococcus. **Infect. Control Hosp. Epidemiol.**

2003;24:362-386.

22. Rihn J.A., Michaels M.G., Harner C.D. Community-acquired methicillin-resistant *Staphylococcus aureus*: an emergent problem in the athletic population. **Am. J. Sports Med. 2005:33:1924-1929.**

23. Tiemersma E.W., Bronzwaer S.L., Lyytikainen O., Degener J.E., Schrijnemakers P., Bruinsma N., Monen J., Witte W., Grundman H., European Antimicrobial Resistance Surveillance System Participants. Methicillin-resistant *Staphylococcus aureus* in Europe, 1999-2002. **Emerg. Infect. Dis. 2004;10:1627-1634.**

24. Centers for Disease Control and Prevention. Methicillin-resistant *Staphylococcus aureus* infections among competitive sports participants- Colorado, Indiana, Pennsylvania, and Los Angeles County, 2000-2003. **MMWR Morb. Mortal. Wkly. Rep. 2003; 52:793-795.**

25. US Centers for Disease Control and Prevention. Methicillin-resistant *Staphylococcus aureus* infections in correctional facilities– Georgia, California and Texas, 2001-2003. **MMWR Morb. Mortal. Wkly. Rep. 2003; 52:992-996.**

26. Vandenesch F., Naimi T., Enright M.C., Lina G., Nimmo G.R., Heffernan H., Liassine N., Bes M., Greenland T., Reverdy M.E., Etienne J. Community-acquired methicillin-resistant *Staphylococcus aureus* carrying Panton-Valentine leukocidin genes: worldwide emergence. **Emerg. Infect. Dis. 2003; 9:978-984.**

27. von Eiff C., Becker K., Machka K., Stammer H., Peters G. Nasal carriage as a source of *Staphylococcus aureus* bacteremia. Study Group. **N. Engl. J. Med. 2001;344:11-16.**

28. Yu V.L., Goetz A., Wagener M., Smith P.B., Rihs J.D., Hanchett J., Zuravleff J.J. *Staphylococcus aureus* carriage and infection in patients on hemodialysis. **N. Engl. J. Med 1986;315:91-96.**

Glossary

Bacteremia. Bacteria growing in the bloodstream as evidenced by blood cultures.

CA-MRSA. Community-acquired or community-associated MRSA.

Cellulitis. An infection of the skin leading to redness, swelling, and pain.

Clindamycin. An oral and intravenous antibiotic effective against MRSA (resistance is possible).

Daptomycin. An intravenous antibiotic effective against MRSA (resistance is possible).

Decortication. A surgical procedure that involves thoracotomy and the cleansing of the pleural space.

Dialysis. A medical procedure to remove wastes or toxins from the blood and adjust fluid and ions (potassium, sodium, bicarbonate or calcium).

Doxycycline. An oral and intravenous antibiotic of the tetracycline group. Effective against MRSA (resistance is possible).

Empyema. A bacterial infection of the pleural space which normally is a space that covers the lungs and is sterile.

Endocarditis. Inflammation or infection of the lining of the heart and/or its valves.

HA-MRSA. Hospital-acquired MRSA.

Leukocidin. A toxin capable of killing or inactivating white blood cells.

Linezolid. An oral and intravenous antibiotic that is active against MRSA (resistance is possible).

Methicillin. A semi-synthetic penicillin, no longer used clinically, but which is still mentioned for historical reasons.

MRSA. Methicillin-resistant *Staphylococcus aureus*. Also called ORSA (oxacillin-resistant *S. aureus*).

MSSA. Methicillin-sensitive *Staphylococcus aureus*. Also called *Staphylococcus aureus*, *S. aureus* or Staph.

Mupirocin. A topical antibacterial active against *S. aureus* and MRSA.

Nafcillin. An intravenous antibiotic active against MSSA but not MRSA.

Necrotizing fasciitis. A severe tissue infection with destruction and inflammation of the fascia (connective tissue under the skin).

Oxacillin. An intravenous antibiotic similar to nafcillin and active against MSSA but not MRSA.

Quinupristin-dalfopristin. A combination of antibiotics effective against Staph and MRSA.

Rifampin. An oral antibiotic active against MRSA. Not recommended to be used alone. It has been used to treat tuberculosis as well.

Scalded skin syndrome. An acute skin disorder seen in infants and immunocompromised individuals due to a toxin from *S. aureus*. It is characterized by diffuse skin redness, necrosis, exfoliation and peeling of the skin.

Staphylococcus aureus. Also called Staph, *S. aureus*, or MSSA.

Toxic shock syndrome. An acute and sometimes fatal disease that is characterized by fever, diarrhea, nausea, diffuse skin redness, and shock caused by *S. aureus*. This syndrome was first observed in menstruating females using tampons.

Trimethoprim-sulfamethoxazole. A synergistic combination of antibiotics, used orally and intravenously against MRSA (resistance is possible).

Tigecycline. An intravenous antibiotic effective against Staph and MRSA.

Vancomycin. An intravenous antibiotic used against Staph and MRSA.

Index

Abscess (es)   17, 21 22

Amoeba   14, 66

Antibiotics   3, 4, 9, 10, 17, 18, 19, 21, 22, 25, 31, 32, 36, 37, 38, 39, 42, 43, 49, 51, 60, 76, 77

Bacteria   1, 2, 3, 7, 9, 10, 16, 20, 30, 38, 59, 61, 73, 75

Bacteremia   2, 16, 70, 73

Cellulitis   15, 73

Chlorhexidine   25, 33, 61

Clindamycin   22, 73

Colonized   1, 2, 3, 4, 5, 9, 20, 25, 27

Contact   3, 7, 9, 13, 14, 26, 28, 59, 60

Contact Isolation   26

Cytokines   16

Daptomycin   4, 22, 73

Decolonization   20, 59, 61

Decortication   41, 73

Dialysis   11, 33, 39, 74

Doxycycline   22, 74

Empyema   40, 74

Endocarditis   17, 74
Handwashing   8, 27, 57
Leukocidin   18, 65, 67, 70, 74
Linezolid   4, 22, 33, 75
Methicillin   iii, 3, 4, 9, 63, 65, 66, 68, 69, 70, 75
Molecular   19
Mupirocin   20, 25, 33, 52, 59, 60, 68, 75
Nafcillin   4, 75, 76
Necrotizing fasciitis   2, 17, 18, 68, 76
Nose (s)   1, 2, 20, 25, 32, 59, 60
Oxacillin   4, 19, 75, 76
Pneumonia   1, 17, 18, 26, 34, 35, 37, 38, 40, 64, 65, 67
Pets   11, 24
Prevention   2, 23, 26, 64, 69
Quinupristin-dalfopristin   4, 22, 76
Rifampin   22, 33, 76
Septic shock   16
Soap   8, 23, 24, 25, 57, 61
Spider   12, 15, 25, 30, 49, 50, 52
Spread   7, 14, 26, 27, 45

Sputum   19, 38, 40

Surveillance   27, 28, 69

Tigecycline   4, 22, 77

Toxins   13, 15, 17, 18, 74

Toxic shock syndrome   1, 2, 17, 77

Transmission   9, 26, 27, 68

Trimethoprim-sulfamethoxazole   22, 33, 77

Vancomycin   4, 22, 31, 77

Vignette   29, 34

www.ingramcontent.com/pod-product-compliance
Lightning Source LLC
Chambersburg PA
CBHW020453220526
45464CB00002B/975